THE BASICS OF RAISING BACKYARD RABBITS

BEGINNER'S GUIDE TO RAISING, FEEDING, BREEDING AND BUTCHERING RABBITS

DAVID NASH

PREFACE

Since you are reading a book on self-reliance, I am assuming you want to know more about how to take care of yourself in disaster situations

I would like to suggest you take a moment and visit my website and YouTube channel for thousands of hours of free content related to basic preparedness concepts

Dave's Homestead Website
https://www.tngun.com

Dave's Homestead YouTube Channel
https://www.youtube.com/tngun

Shepherd Publishing
https://www.shepherdpublish.com

1

WHY KEEP RABBITS

For many years my Paternal Grandfather raised rabbits and quail commercially on the gulf coast of Louisiana. I always wanted to try my hand at it on a smaller scale. My reasons for wanting rabbits are deeper than pure sentiment.

Number one is my quest for greater self-sufficiency. While no one can be totally independent in today's America, Rabbits are an easily grown source of protein.

I have read many stories from Great Depression I where children gathered grasses on the way home from school to feed their families meat supply.

Rabbit Meat is Sustainable

Rabbit meat is sustainable. Which, I use the term to mean that my rabbit meat is produced in a way that is healthy for consumers and is a humane, ecologically harmless, socially responsible and economically feasible source of protein. That's a wide range of criteria.

While I am no eco-fascist I believe that when God gave man dominion over the earth, he entrusted us with the responsibility to use the resources he entrusted us with in a manner that best serves our needs while preserving his bounty. I can feed my rabbits vegetation from my garden, as well as forage I can harvest feed from the lot behind my home.

The Meat is Healthy

Rabbit meat is healthy. It is extremely lean and has little cholesterol. I have an abundance of food stored near my midsection, and can use some extremely lean meat. Early arctic explorers learned of "rabbit starvation" where it is possible to starve to death if you eat a diet of high protein lean meat without any fat. Being able to raise my own meat without hormones and without the high fat of my normal food will make it that much easier for my goal of "getting healthy" (my wife does not let me use the word diet…)

Great Fertilizer

Rabbit feces is the only animal feces that can be used directly on plants without "burning them" No composting required. I have heard of serious gardeners buying this byproduct to use on their gardens. Myself, I use the waste to raise worms. I use the worms to supplement food for my chickens (and fish as soon as I get the gumption to dig the pond).

Rabbits Breed Fast and Eat Cheap Food

Rabbit breed fast. Their gestation period is 28-35 days. And the rabbits reach market weight (four to five pounds dressed) in about 8 weeks. They also have a high feed to meat conversion factor. According to Vet Scan, they convert aver 20% of their feed into meat. This coupled with a high meat to bone ratio means you get a lot of meat for the cost of feed.

Basically, rabbits need fresh water, 1 ¼ cups of food a day, protection from the elements (very heat sensitive) and basic care.

I have mounted a set of 4 cages with built in nesting boxes in my carport. These cages have a water trough attached that I can easily refill and clean out as necessary. Being this close to the house, watching them is easy. However, being this close to the house cleanliness is a BIG concern. I made a composting bed of 8ft 2×4's and 3mil thick plastic. But this is a catch basin for anything that doesn't hit the plastic tubs I set in the bed. Once I got a couple inches of waste in the tubs, I will seeded them with a few handfuls of red worms.

You Don't Need a Lot of Space

You don't need a 40 acre farm to keep rabbits. I live in a small suburban lot, with less than 1/3 acre of yard. With a good pen and quality feed you can keep a couple of rabbit breeding trios very satisfied and healthy with room to spare.

I keep a trio of rabbits, which is two does and an unrelated buck. This provides the ability to breed all the rabbit my family needs.

2

10 BEST BACKYARD RABBIT BREEDS

Meat, Fur, or Show?

There are three main types of rabbits, Fur rabbits that can be used to gather wool, show rabbits to win ribbons, and meat rabbits used to feed your family.

I know nothing about she rabbits, and very little about fur rabbits. Luckily this book is not about them, keeping rabbits for reasons other than meat are outside the scope of this work.

Below are some rabbit breeds that are best suited for beginner keepers of meat Rabbits.

American Chinchilla:

While I have not raised this particular breed, many consider it to be one of the best meat rabbit breeds. It looks like a large chinchilla rodent.

American Chinchillas have a stocky body and the meat may weigh up to 9 pounds. They are great for roasting and barbecue.

Californian:

Like most backyard raisers of meat rabbits, most of my experience is with Californians and New Zealand whites.

The Californian green was developed by crossing of Chinchilla with New Zealand White. They have white fur with black spots and are known for their blocky and good production of meat. They can weigh around 8 to 12 pounds. This is a very easy rabbit to find if you are looking for new stock.

Champagne D Argent:

This breed has been used as a meat rabbit since the early 1600s. It is a popular meat rabbit worldwide. It can weight up to 12 pounds and is also a common pet rabbit breed.

They are available in white, creme and chocolate colors.

Cinnamon:

This breed was "accidentally" created in 1962 by crossbreeding between the New Zealand and American Chinchilla. Having sturdy body of both of the breeds, this rabbit produces a good quantity of meat and are bred commercially for that purpose. It is also a recognized show breed in America and is often kept as a pet or for fur. It is considered an "all purpose breed".

Flemish Giant:

The flemish giant is considered to be the largest breed of the species. They may get up to 22 pounds although 15 is normal. They are very docile and are kept as pets, but they are commonly raised for meat and fur so they are considered a utility breed.

Due to their size, they consume more feed, wire cages can damage their feet, and they need at least 5 square feet of cage space.

Since Flemish Giants do not reach full size until they are 1.5 years old I do not consider them to be a good homestead rabbit personally, but others may disagree.

New Zealand White:

Not only is the New Zealand White one of the most common rabbit breeds in the US, it is the breed I have the most experience with. They are bred for meat, pelts, show, and laboratory uses. They have the distinction of being the most commonly used breed of rabbit for testing and meat production. The meat can weigh 9 to 12 pounds and fryer rabbits are ready for slaughter at 8-12 weeks.

Palomino:

Palomino rabbits have a smaller bone structure than other meat rabbit varieties, this means they have a good meat to bone ratio. This breed of rabbit is hardy and docile

Palomino rabbits are very clean and they will groom themselves as well as each other. The does are known for being excellent mothers. The fryers grow quickly and reach slaughtering age earlier.

Rex:

Rex rabbits are fur and show rabbits, but also are kept for meat production. It is the number one breed for fur production.

Fur as a by product of meat production is not as good quality because the goal for meat rabbits is the harvest of young rabbits, where it takes time to produce a quality adult pelt.

When mature, they may weigh around 8 to 10 pounds and may come in a variety of blue, amber and spotted patterns ion their color.

Satin:

These are primarily show animals, but are excellent mothers and good for producing meat. These breed are also very rare and may weigh 10 to 12 pounds.

These medium large sized Rabbits are available in blue, black, copper, chocolate, red, Siamese and otter colors.

Silver Fox:

This rabbit is on the livestock conservatory list as a threatened breed. It was developed for meat and fur production.

It can dress out up to 65% of its live weight. The does have large litters, produce plenty of milk, and are excellent mothers. Something unique with silver foxes (at least in my experience) is that they make wonderful foster mothers. If you raise rabbits you will one day see a mother eating their newborn kits, and will want to have rabbits that are good mothers after that sight.

Silver Fox are known for their docile and gentle nature.

Bonus Breed

Altex:

1994 the University of Alabama worked with Texas A&M to develop the Altex meat rabbit. It was developed from Flemish Giant, Champagne d'Argent, and Californian stock, and later with New Zealand White crossings.

Altex rabbits weigh 10 to 20 pounds and have coat markings similar to the Californian rabbit.

3

RABBIT HUSBANDRY

Of all my livestock, I have had the best luck with rabbits. I have done it the longest, and have harvested enough that I think that I have a pretty good system in place. I am by no means an expert, but I do have the basics down.

As with all things I am doing on my urban homestead, I place great importance on having good resources to find answers to my questions. I have a large library of rabbit raising books that cover a lot more than this basic guide.

The first thing is that when you want to raise rabbits, you should have a plan for what you're going to do with them. If you plan on eating them you need to schedule the breeding, weaning, and growing out stages so you don't spend more on feed than you are getting in meat. Creating this system is what has taken me the longest to accomplish.

The way we maximize our outputs while not overworking the does or creating so much rabbit meat that we are overwhelmed is by having three does. One is always pregnant, one is always nursing, and the third is resting.

We have a four cage hutch, and the male is one the far left side. When we want to breed, we take a female out of her cage, and introduce her to the male cage. We tried the other way, but the male wants to sniff around and the female is protective of her territory. If we put the female in with the male he jumps right to work.

I have left the doe in overnight, but I find that it causes fewer problems if I just watch them and then remove the doe to her cage after breeding. I then put her back with the buck about 8 hours later so he can have one more chance at breeding. After that second breeding I don't let them breed again as rabbits ovulate by opportunity rather than on a cycle so it is possible for the female to be breed twice and have two sets of fetus at different developmental stages. This can cause medical problems.

You can check for pregnancy after 10 days to 2 weeks, you can palpate the lower abdomen of the doe with your thumb and forefinger checking for nodules about the size of a marble. It's kind of difficult to do, as it takes practice to get a feel for what is going on, but it does not hurt the doe or the babies as long as your not overly rough.

29 days after breeding, you should put in the nest box. My cages have a built in nest box so I just give them some straw. If I give it to early, they either soil it or eat it. 31 days after breeding she should give birth which in rabbits is called kindling.

Usually kindling takes about 10 minutes or so, and it's normally done at night. I have never had a rabbit give birth when I was around. They like their privacy. Last winter I had a hard time with the litters surviving as it was very cold and it was the doe's first litters so they did not pull enough of their fur to keep the babies warm. Now that the mother's have had a couple litters a piece, they now pull enough fur, and a lot more of my baby rabbits are surviving. The babies are born helpless, furless, and blind.

If they don't have a box to keep them confined with their litter-mates they don't have enough body heat to keep from freezing. It does not matter where the babies are, the mother won't move them, so if you don't have a nest box to keep the babies confined, and one crawls away and starts to freeze, the mother will just let it happen.

I try not to disturb the mother or move the babies for at least three days after birth, but I refrain from the temptation of handling them unless I have a valid reason. The doe will only nurse once a day, and she doesn't like doing that if it is noisy. In the wild, rabbit mother's have an instinct to run away if threatened so that predators will follow her AWAY from the nest. So you may think momma is not nursing, but she probably is.

In about a week (more like 9 days) the babies will have grown their fur. By the 14th day, they will have opened their eyes and will be hopping around.

After they start to move around, if it's warm you should remove the nest box, but even if its cold, you should remove it by three weeks, as the babies will soil it with their wastes and it could cause infections to spread.

After about 4 weeks I begin to move the babies to my grow-out cage. I do this one a day, so that the mother will gradually reduce her milk production.

I keep the rabbits in the grow-out cage for 10 – 12 weeks. At that point they aren't likely to fight, but if you keep them longer, you will need to separate them. I like butchering them as soon as they weigh about 10 pounds. That means I will get about 6 pounds of meat after they are dressed.

Keeping rabbits is pretty simple, keep the cages clean, their water fresh, and don't overfeed them and you won't have too many problems that you cannot find the solutions for. The meat is tasty and very wholesome, but I have to warn you 4 rabbits quickly turn to 40, and its easy to spend a lot more on raising rabbits than you would just spend buying meat at the grocery. However, grocery bought meat cannot compare to the freshness and satisfaction from growing your own.

4

HOW TO BOTTLE FEED BABY RABBITS

Unless the mother rabbit is dead, she is probably feeding her babies, even if you think she is ignoring them. Rabbit does feed their babies twice per day. This is normal and natural: in the wild, a mother rabbit not in the process of feeding her offspring stays as far away from the nest as possible to avoid attracting predators to her babies.

If you're messing with the babies too much she won't take care of them. It is her natural instinct to lead predators away from the nest so she will NEVER nurse or care for her babies if she sees you as handling them.

Things to Think About Before "Helping"

Sometimes mama rabbit seems to be "ignoring" her litter, this can be a way to keep predators away from the nest. Before you check the babies understand that you may be interfering with the natural process.

If their bellies are round and full-looking, they're warm, their skin is a healthy dark pink, and not too wrinkled, and they are sleeping, then mama is feeding them.

However, if the babies are very wrinkled, cold, bluish in color, have shrunken bellies, and perhaps are even crawling around looking for food, then you may have to feed them.

If the babies really need to be fed, here is how we go about bottle feeding baby rabbits:

Equipment

Nursing syringes

- We get ours from the local co-op, but you can order them online.

KMR (Kitten Milk Replacer) heated gently to about 105degrees Fahrenheit

Procedure

- Without a doubt the most important thing to avoid is aspiration (inhalation) of formula by the babies. The smallest drop of formula in the lungs can cause fatal pneumonia or drowning. This happens more often than not.
- Feed the baby rabbit very small drops, just touched to the lips. The kit should try to lick the milk from its mouth. Be prepared for the kit to move around when its mouth is touched. Have also a paper towel or absorbent cloth at hand to mop any extra milk or fluid that runs up into the nostrils. Blot often!
- Feed the kit very small amounts until it gets the taste of what you are giving. At this point they will often demand more … don't give it to them! Keep the small amounts going as long as the kit will take them … then give it a break of fifteen minutes to half an hour and do some more.
- When the belly is tense, they've had enough until they

urinate… or until the tenseness vanishes and they are showing signs of hunger again. Since most nursing formulas are lower in nutrient content than doe milk, it is important to keep the kit full and well-hydrated.

- When you are done feeding it helps to stimulate their bodies to get rid of waste. Normally the momma doe would lick their bellies until they void themselves, what we do is to gently rub a finger down their belly. I have heard that if they have litter mates, the natural jostling they give each other will do the same thing.

In Case of Accidental Aspiration

If the baby aspirates formula, it will completely block the airway and cause the baby to pass out. This is normally a death sentence, but the following "Bunny Heimlich" maneuver is the only hope of saving the little one.

• Hold the baby very firmly between your palms, one on each side of the rabbit

• Stabilize the back and neck firmly so they do not move at all, raise the baby above your head, so his nose is pointing up.

• Use a firm, downward motion to swing the kit towards your feet. Your trying to use centrifugal force to expel the thick formula, but not trying to recreate the G-Force from a fighter plane. Don't go too fast or too violently.

• Repeat the procedure two or three times, as necessary. The weight of the baby's internal organs pressing against the diaphragm when you swing downwards usually provides enough pressure to expel air from the lungs, as well as the drop of milk blocking the airway.

• Once you feel the baby begin to move, STOP IMMEDIATELY.

• Consult with your veterinarian about whether or not to place the baby on prophylactic antibiotics to prevent aspiration pneumonia.

The Reality of Keeping Meat Rabbits

I really don't want to sound harsh or uncaring, but the reality is that if the mother doesn't feed the babies, and you do not have a lactating doe to feed the newborn kit, the rabbit will, in all likelihood, die. Even with feeding, the KMR doesn't have the colostrum that is needed to ensure the new rabbit has a healthy immune system.

This technique works, but generally I don't attempt it anymore.

Instead, what I do is keep records of the quality of the mother. A doe gets two chances to raise a strong litter or I harvest her.

As far as eating her babies, she gets one chance, as there may have been a reason, but a second time calls for the stew pot ad there is no cure for a doe that constantly eats her young.

The truth is that rabbits are prey animals and are driven to procreate. As such you can breed characteristics into, or out of , your personal rabbitry. I want rabbits that do not bite or scratch, and are good mothers. Rabbits that conform to those ideals get the chance to spread their genes. Those that do not get harvested.

5

BUILDING SHELTER FOR YOUR RABBITS

There are a couple of ways to keep rabbits, however, they all have some drawbacks.

Most people that keep rabbits use some type of wire cage with a cage floor to allow the rabbit pellets and urine to fall freely.

If you have a solid floor, you WILL have problems with waste.

However, if you have a heavy breed of rabbit, wire cage floors can be painful to their feet.

I use a plastic "resting board" in all my cages, it is a small plastic board, that is just a little bigger than the rabbit. It has slots to allow the manure to fall free, but the material between the slats is much thicker than cage wire.

Rabbit Cage Requirements

The floor of a wire cage should be made of 0.5-by-1 inch mesh. The top and sides can be made of bigger 1-by-2 inch mesh for the top and side.

Hanging cages in single layers the cages hanging from a ceiling is very easy to manage, if you have double cages to save space you will need to have some manner of diverting waste so the top animal does not soil the animal in the bottom cage.

Adult does and bucks should keep in individual cages. Remember, rabbits ovulate and breed by opportunity. If you keep bucks and does together they will breed uncontrollably. I do keep my weaned rabbits communally in a single grow out cage, as I harvest them as a group before they have the opportunity to breed more rabbits.

Single cages should be at least 30 inches deep, 20 inches high and 30 inches wide. This measurement can be different according to the breeds. The USDA recommends that rabbits have at least 3/4 square foot of floor space per pound of body weight.

My rabbit cages have a smaller recessed nesting box hanging from the larger cage. This is because baby rabbits will crawl away from the nest. By having the nesting box below the regular floor of the cage, kits cannot wander into the next cage and get killed by the other Doe. Alternatively, most people add “baby saver” wire around the bottom 3-4 inches of the cage sides. Baby saver wire is welded wire with smaller holes so that the newborns cannot crawl through the large holes.

In the outside of the cage every cage should have attached hopper of feed and a watering system. Keep some straw or this types of materials in the cages of doe during their delivery time and winter season.

Build or Buy

One would think that building your own cage would save money, but in my experience, the equipment and the time investment required just doesn’t make sense long term unless you are considering a commercial farm.

I have found that there are so many people entering and leaving rabbit raising that you can find good deals on used cages if you take the time to look.

Building a rabbit cage for scratch is our of the scope of this beginner's guide as it could be its own guide by itself, the only real advise I would give you is that when you are planning your own cage, take into consideration the dimensions of your wire. Cutting 2 or three inches off a roll of wire to make a custom cage size gets old really fast. I learned that the hard way attempting to build my own cages.

Examples of rabbit cages and hutches from my backyard rabbitry

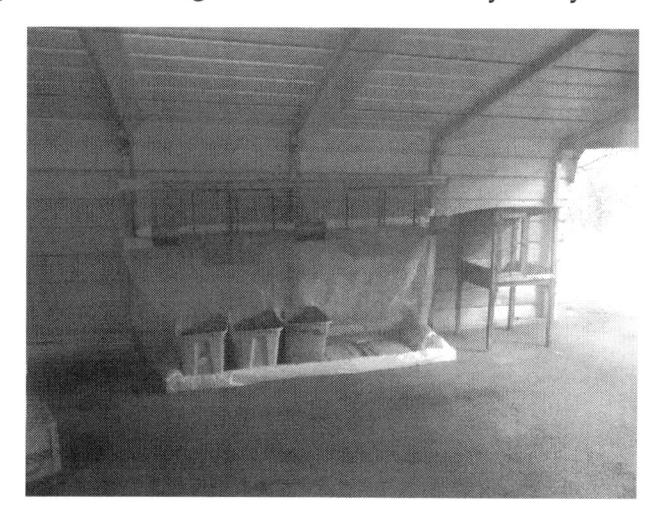

Here is an early example of a rabbit shelter setup at my own house

Notice that I have a means to collect waste, an automatic watering system, a grow-out cage, protection from the weather, and plenty of ventilation.

I really liked this setup, but my wife did not like rabbits in her car port. Apparently, I had a buck that did not like her, and he would try to urinate on my wife when she parked in the carport.

Billboard Tarp Cage Cover

She made me move the cages to the backyard, where I mounted them on my fence. At that point I build ta larger grow-out cage, which you can see on the left of the image above. The base of the cage was a salvaged crate that I removed the bottom, front, and top. I hinged the top so I could life it up. I replaced the bottom and front with rabbit cage wire that you can purchase by the foot at a local co-op or feed store.

The top was built from grey plastic conduit pipe bent into a rounded shape and anchored to fence posts. The cover is a thick recycled bill-board tarp. This was a very cheap setup, and lasted for years until I decided to take a break from keeping rabbits.

Latex Cement Hutch

My current project is in the works, it is a hutch using latex cement as a building material. I have been experimenting with latex cement as a homestead building material, and am very pleased so far. This is a frame built out of 2x4 and 2x3 lumber and then had cheap goodwill sheets and curtains stapled to it. I then painted the sheets with a slurry made up of PVA mis-mixed paint from the hardware store, water and type S mortar. I talk more about latex cement in How I Built a Ferrocement "Boulder Bunker" which is available on Amazon in electronic, print and audio format. The audio format comes with an addendum file of instructions. The unpainted sheet will eventually be molded and covered with latex cement to create a formed "chute" which will allow the waste to fall into a collection bucket for sale to gardeners.

Rabbit Colonies

Ancient Roman's kept their rabbits in colonies called Leporaria, which were stone walled open pens where rabbits were allowed to freely breed in unsupervised groups. When a Roman wanted dinner, they just grabbed a rabbit from the open top pen.

Finding this to sound simpler then keeping up with breeding dates I tried it. First I used my old green chicken tractor as shown in The Basics of Raising Backyard Chickens.

My problem was that rabbits dig. They soon, and I mean in less than a week, dug tunnels in and around the tractor and promptly escaped. I stapled chicken wire under the tractor, but it was thin enough they bit through it, and it rusted easily causing a tangled sharp mess. I went back with welded wire goat fencing with 4x4 inch holes. This worked, but it was an added expense and weight on the tractor.

Next I dug down and created a rabbit proof barrier about 8 inches deep and set a dog kennel over the barrier.

This worked well, but it was hard to keep my rabbits from escaping when I went in the kennel to capture one for supper.

Some things to consider for Colony raising:

- It takes more space than cage raising, provide *at least* 10 square feet per adult.
- Protect your rabbits from weather and predators, as you can see I used tarps for wind and rain protection
- Use baby saver wire on the bottom
- Keep things clean with litter boxes and deep litter (or add Diatomaceous Earth to the feed and sprinkle it on the ground - this is what I do and I don't have problems with worms)
- Provide several locations for feeding to reduce competition, they are territorial.
- Consider feeding in areas that can be shut off to help catch the rabbits as needed
- Provide lots of nesting places and materials for the does, with mixed bucks and does, litters are unpredictably timed.
- Have a good amount of grow out cage ready, because you will get lots of litters

- It's a good idea to sex your grow outs and separate when they are around 8 weeks old (alternatively, I just send them all to freezer camp)
- Rabbits in groups are more active, consider providing multiple levels for jumping

6

FEEDING AND WATERING

What Should You Feed Your Rabbits?

I find that rabbits are relatively inexpensive to feed, especially if you are only raising 3 does and 1 buck for a source of home-raised backyard meat.

In my experience, commercial pellets are fine and are easy to find locally. In a pinch rabbits will eat almost any vegetation, bark, grass, trimmings, all things found in the wild, however a wild rabbit is much smaller than a home raised meat rabbit and are not pumping out litters every few months. Lactating does need a 18% protein feed to support adequate milk production.

I try to feed mine a good quality feed because I find that it is cheaper in the long run to get rabbits to a marketable rate faster and dispatch them to save on feed rather than to feed a cheaper quality of feed. I also supplement with a salt/mineral lick, but many brands of pellets contain minerals so no salt/mineral lick is needed.

Check to see if your pellet mix contains Copper Sulphate. This mineral helps the rabbit fight off intestinal parasites, which is especially important if you are raising in a ground based colony.

You can, and I do on occasion, supplement with whole oats, grass hay, dried greens fruit and vine trimmings, as well as an occasional fruit treat. At the end of this chapter I talk about a treat I give my rabbits that I make out of the leftover pulp from juicing vegetables. My rabbits love that!

I buy my feed at the feed store in 50 pound sacks, pet store feed costs just as much for a five pound bag.

Once you find a good pellet, always use the same brand and type of pellets. If you must change, mix the old feed with the new feed over 2 or 3 days. Some rabbits do not do well to the sudden change in feed and could cause digestive problems. When you buy a rabbit from a breeder, or if you sell a rabbit to someone, the purchase should include a small amount of the current food.

Attached individual cage feeders are around $6.00 online

I also like to supplement the pellets with hay, to keep the rabbit from pooping on it I use a hay feeder. Hay is high in fiber to help digestion. But you want to avoid feeding your rabbits straight alfalfa hay. Alfalfa contains a comparatively high amount of calcium. High calcium levels can cause kidney stones.

How Much Feed and Water per Rabbit

The amount that rabbits are fed depends on your rabbits and the conditions you keep them in. They need more food in cold weather and less in hot. You should watch your rabbits and feed according to how lean or how fat they are.

You need to know what a healthy rabbit looks and feels like. You need to limit the feed for full grown Bucks or non-breeding, fat rabbits decreases fertility

Adult rabbits will eat about four ounces of pellets a day, and does with young need about twice that.

For a meat breed, about 1/2 to 1 cup a day.

Water

The normal rate of water consumption is approximately 4-6 fluid ounces per 2 pounds of body weight daily.

Meaning, expect a 4 pound rabbit to drink one cup of water a day, or less.

Personally, I use automated systems to allow the rabbit to get as much water as they want to drink.

They can be destructive, so I use 2 litter soda bottle upside down with a screw on water nozzle

This type of bottle conversion kit can be found online for around $5.00

They tend to freeze up in the winter, so I prefer a larger system like the bucket waterer below.

How to Build a Rabbit Bucket Waterer

This following is a very simple watering device that makes my daily chores much easier.

poultry nipples are little fittings that you screw into a bucket or a pipe and when an animal pushes it, a little water comes out. This makes a very simple chicken waterer that makes my daily chores much easier.

Poultry nipples can be found online

I bought a set of the nipples from an online store, and screwed them into a 5-gallon bucket I got from a local bakery. It is important to use food grade buckets, as some manufacturers use release agents that are not healthy.

Once I got everything adjusted right this thing works awesome…

One the years I have uses poultry nipples to make all manner of waterers. The simplest is simply a bucket with a few nipples screwed into the bottom. However, I have also had good success with more complicated set ups using PVC pipe and nipples screwed into a large diameter pipe that is filled from an elevated bucket.

I have also known people to make an automatically filling setup by plumbing in a toilet bowl float to keep the system full.

It works for almost any caged animals. I also use this system with my chickens.

It does not need to be complicated.

Buy some poultry nipples, find a food grade bucket, devise a method of hanging the bucket high enough the chickens can get under it and peck the nipples.

Drill 4 or 5 holes spaced equally apart with a 11/32 bit.

If you want to get fancy, a heat gun may make threading the nipple into the bucket easier, but do not go overboard and melt the plastic.

Next simply thread in the nipples and invert the bucket.

Hang it do that your animals can easily get under it to drink without being cramped.

Fill with water and check for leaks.

I find that I can go about three days before I need to refill. Sometimes I still have water, but I like to keep clean water in the bucket so it does not grow bacteria or algae.

I make a point to check the water as I feed my rabbits so it is not an issue.

How to Make Rabbit Treat Sticks from Leftover Juicer Pulp

I have a friend that swears by juicing, and I was intrigued by it, so when I saw a new Big Boss juicer on sale for 2/3 its normal price I decided to buy one.

The juicer is very easy to use, and I prefer drinking my vegetables to eating them, but I hate to waste all the good fiber that is left in the juicer.

I have seen recipes for veggie burgers and other food items from the remains, and I plan on trying them later.

However, what I wanted to try first was using the leftover bits to feed my rabbits. Below is my simple procedure to turn leftover juicer pulp into rabbit treat sticks.

The procedure to make Rabbit Treats from Juicer Pulp is very simple:

- When cleaning out the juicer, I collect all the vegetable matter (which is pretty dry).
- Next I press it into shape and dehydrate it.
- It makes a dry stick (and if I had a pellet machine it would be VERY close to how they make commercial rabbit food).
- I took it out to my rabbits, and they loved it.

I get to eat my vegetables how I prefer, and the rabbits get a health treat to munch on win-win…

7

RABBIT TATTOOING

I raise plain New Zealand White meat rabbits, and even though I mess with them daily, I cannot always tell them apart. Most days it doesn't make much of a difference. However, trying to keep records can be hard without it. Keeping track of all the important things a responsible breeder needs to know like; who has had medication and who is due for breeding, is much easier if they were easily identified. I found a solution for this from the show side of rabbit production. (I also made a paper notebook that contains forms to help with my record keeping, you can find Rabbitry Records on Amazon.

Required for Showing

The American Rabbit Breeders Association show rules mandate that a clear and readable tattoo be present in every rabbit in a sanctioned show. This is important to be able to prevent lowlifes from stealing your rabbit. Or switching their looser rabbit for the champion rabbit you spent so much time on.

Personally I have no interest in showing rabbits. However, I am not going to belittle someone for keeping a rabbit for a different purpose than mine. Especially when we both agree on much of the basics.

If you never plan on showing your rabbit, it does not matter which ear you choose to tattoo. Know that if you ever show your animal, identifying tattoos need to be in the left ear. This is because championship identifiers are tattooed in the right ear. Personally I stick with this standard because it is easy. All my rabbits are marked in the same place.

Keeps Track for Breeding

Some breeders use a complex system where one letter is for the father and the next for the mother with additional numbers for litters. Others have a system where the numbers are meaningless and are only identifiers. The possibilities are endless. By developing a unique system, it is possible to ascertain the animals identity in great detail.

Personally for my set up My male is number 1 and is kept in the leftmost cage. The female to his right is #2, the next #3, and the doe on the right most side is #4. I don't tattoo my babies as they are going to be eaten at about the time they get large enough to tattoo. When I replace a breeder rabbit, the replacement gets the number of the rabbit it is replacing....

The way it works is not unlike primitive tattooing on humans. The skin of the ear is penetrated with needles. Ink is rubbed into the holes, and once healed leaves a permanent mark.

To make a legible mark and to reduce the stress on the animal as much as possible a special tattoo clamp has been designed. Basically it is a large set of pliers with one side being a rubber backer, and the other leg holding special alpha-numeric needle block. When you place the clamp on the ear and squeeze it punches the ear to allow inking.

Place in Left Ear

Besides tattoo placement in the LEFT ear, there are some other placement considerations you need to be aware of. The tattoo should be deep enough in the ear to avoid any hair around the fringe of the ear. This is because this hair will often impair the legibility of the mark. You also need to take care to avoid the vein that runs down the ear as well as any other large blood vessels.

Shine a strong flashlight through the ear and you can see them. If you accidentally puncture one of the vessels it may cause above average bleeding. However, but it should not cause any long term harm to the animal. Usually it will stop bleeding in a short time. Place the tattoo clamp correctly and you will have little or no bleeding. Proper tattoo inks contain alcohol to aid in healing and help prevent infection should some bleeding occur. Do not rely on the alcohol in the ink to take the place of proper clinical principles.

Clean and Confirm

Clean the area with an alcohol swab to remove any oil and dirt that is in the ear. This will not only assure a clean site to insure fast healing, but also helps make the final result a clearer mark.

The tattoo should be placed in the ear so that it is readable when the ear is opened and looked at from the left side of the animal. By positioning the animal with the nose to your left, you will insure that the mark will be readable from that position when finished.

Many rabbit tattoo sets will only allow the digits to be placed into the tool "right side up". You will not be able to make the mark that is upside down when tattooing from the left side of the animal.

Confirm everything is right by testing the mark on a piece of paper prior to tattooing the animal. Do not forget that the digits will appear reversed while in the tool. Be careful that you have the digits in the proper order.

Restrain and Pierce

Restrain the animal to prevent injury to both the animal and the operator should the animal move suddenly during the tattoo process. This is also important to assure that the final result is desirable as any sudden movement while tattooing the animal will result in an illegible mark. Many people use boxes, but just wrap the animal in an old towel and have a helper hold the rabbit down.

Some build a small box that has a hole in the top that the ear can stick through. If tattooing several rabbits in one session, this idea might prove useful and speed up the process.

Realize that the animal will most likely try to move as you squeeze the tongs, and be ready for it so you cause as little pain as possible and get a clear legible tattoo.

After positioning the tool properly, squeeze the tattoo tongs together firmly. A common mistake is to squeeze the tongs until the rabbit responds and then release the tongs. This will stop the needle before the digits have thrust through the inner skin of the ear. It is imperative that the tongs be completely closed to make a penetrating imprint. On young animals it is not uncommon for the digits to penetrate completely through the ear. After gaining experience, you will develop a feel for the right pressure to apply to the tongs to make a good mark. Don't be alarmed if the needles penetrate the ear. This will still result in a legible tattoo and the backside of the ear will heal over and not leave a noticeable mark on the outside. It is better to go completely through the ear than to apply too little pressure and leave a mark that is illegible.

Rub in Ink and Let Heal

After making the imprint, apply ink to the tattoo with a brush. Rub the ink vigorously into all the puncture holes. (Tip: The bristle brush can be trimmed shorter with scissors to make a stiffer brush for better penetration.)

If the tattoo needs to be read within a short time of the tattoo process, you may wipe the excess ink from the ear with cotton or tissue and then apply a light film of Vaseline to the tattoo. It should be distinct and legible immediately. If you do not need to recognize the tattoo immediately, it is not necessary to wipe the excess ink away. Personally, I let the excess ink wear off naturally.

This is a very simple process, and the costs are pretty reasonable. I believe my kit was around $30.00 and was easily found by doing a simple internet search.

8

SELLING RABBITS

Selling live rabbits is relatively unregulated. I often go to the local livestock auctions to see what they are selling for, but I typically don't buy there as most of the time it is a farmer getting rid of a problem animal.

Selling butchered rabbit is a different story. Typically that is illegal without expensive licenses, inspectors, and specialized buildings.

However, while outside the scope of a beginner guide, there is an exception it would pay for the small scale farmer to understand.

The US Department of Agriculture made an exemption for small-scale on-farm poultry producers that does not require poultry to be harvested and processed at a USDA inspected facility; it is commonly referred to as the 'poultry exemption.' (While it says poultry exemption it includes rabbits)

While you should research this yourself, and contact your state agriculture department if you decide to sell your rabbits on an ongoing basis, the framework is as such:

If you sell less than 1000 processed Poultry (and rabbits) a year, and you sell direct to consumer only (no wholesale/sold to be resold) you can avoid the expensive licensed processors.

Poultry (and rabbits) must be raised and slaughtered on site (cannot purchase/process poultry or rabbits raised by others)

Also, the following restrictions apply (for poultry and rabbits)

• If no employees are used (processed only by immediate family members), the meat can be sold on or off the farm (i.e. farmers market).

• If employees are used, the meat can *only be sold on the farm.*

9

USEFUL BY PRODUCTS

Rabbit Manure

Fresh rabbit manure is approximately 2 percent nitrogen, 1 percent phosphorus and 1 percent potassium. What makes it a marketable product is that it can me used fresh, straight from under the hutch. It does not burn plants like other manures.

I have had success in selling 5 gallon buckets of rabbit manure to gardeners.

I am sure you can figure out how to market this, an internet search showed a commercial 10 pound bag of rabbit manure selling on Amazon for $59.0

Earthworms

I have also used the manure to create a bed for earthworms. Whenever you talk about raising rabbits commercially, you will heard stories of people making money selling earthworms that grow in the manure beds under the rabbit cages.

Personally, I have raised a lot of earthworms in this manner, but I normally use them to feed my chickens rather than sell. I am sure it can be done, and there are a lot of information on vermicomposting (raising worms in garbage) but I have never sold any. I just haven't tried.

Fur

Rabbit fur is not something that is raised commercially in North America. As mentioned earlier, the pelts of meat rabbits generally are not old enough to be of a quality to be used in the fur trade.

There is a bit of a market for rabbit wool, especially from rabbits like Angora that are raised to be shorn of their fur and have it turned into yarn. A few years ago while I was recuperating from surgery I experimented with a homemade loom and a drop spindle to make yard and weave.

In the research for this I learned how easy it was to make felt, and the small artisan market for the materials.

As a rabbit keeper, you will end up with fur all over the cages, it should be possible (although I have not tried it myself) to harvest that waste and turn it into felt.

Making Felt

Felt making is making fabric by locking together fibers (generally wool) using friction and moisture. This is the traditional 'wet' felting technique, although now there is also needle felting, which doesn't require moisture or friction. Felt is one of the world's oldest textiles (if not *the* oldest).

There are different ways to make felt, most often home welters use wet felting or needle felting. I have no experience with needle felting so I won't attempt to describe it,

To wet felt, you basically lay tufts of fur in a single direction (example North to South) over some kind of mat. Most online tutorials say bubble wrap, but I always bust those so I use parchment paper.

Once you have the size you want orientated in one direction, create a new layer orientated 90 degrees from the original (i.e. East to West)

Make a third layer with the tufts facing the original direction (N-S)

Cover with tulle netting and sprinkle with water.

Push down the lasagna pile of fur until the rabbit wool soaks up the water.

Hold the net down with one hand and rub a bar of soap gently through the net and into the fur mat.

Gently work the soap in circular motions, what is happening is the tiny hooks in the individual hairs are catching on each other like hook and look fastening. Eventually the loose mat of fur will become a thick mat.

When that happens, peel of the net from the wool, and the wool from the mat.

Place the wool sheet onto a sushi mat and roll. Tightly to drain off water.

Now role the rolled tube of mat back and forth on a table for a minute

Unroll and peel the felt of and turn 90 degrees, re-roll the bat back into a tight tube, and roll the tube against a hard surface.

Repeat this rolling and turning until you can pinch the felt and the layers won't flake off.

Rinse the felt in water, but finish the rinsing in cold water.

Once again, I am sure there is someone willing to buy this felt, but its probably on Etsy. I like knowing I can, but I do not think I will become rich doing it.

Tanning Fur

There are many ways to tan an animal hide, and you can even buy a commercial tanning solution from Amazon. But since the commercial stuff is $16 for 11 ounces I wanted a cheaper method. I want to collect enough tanned hides that I can cut the hides into strips and weave a blanket from rolled up strips of fur on rabbit hide to make a fluffy warm throw for my bed. I am having problems with this as my kind hearted wife can't get over the fact that the hides in the freezer used to me her rabbits.

An inexpensive way to see if you can find a marketable use for rabbit hides in by tanning using eggs.

Here is a simple rundown of the process:

1. Skin the rabbit (how to do that is covered in Chapter 11)

2. Stretch out the hide and scrape off any fat and tissues using the side of a metal spoon. With larger animals this fleshing is vital to prevent spoilage, but when I pull the rabbit hide off by hand, I find that very little flesh is left. However, 30 minutes or so scaring with the spoon to remove any membranes left on the hide is invaluable so the hide doesn't get "crusty" with dried flesh.

3. Salt the hide to prevent bacterial growth. Lay out the skin, fur side down, and spread non-iodized salt liberally over the whole thing, making sure to cover the entire hide. 4. Let the hide dry for a couple days, or until it stays flat when you pick it up. You don't have to be in a hurry for the next steps, as a properly dried and salted hide will stay preserved for months as long as it stays dry.

5. Scrape off the salt and soak the hide in fresh, warm water. Do not take the skin from the water until it is soft everywhere.

6. Squeeze the water out of the hide and stretch it in every direction. The more you stretch the more supple your finished product will be.

7. Mix a solution of 1 egg yolk and a half cup of water per hide

8. Rub the egg yolk solution into the softened hide. 9. Work the hide by stretching it in all directions. Keep doing this until the egg is pulled into the hide and it dries. This will take time, a few hours, but if you stop before the egg is completely absorbed into the hide the resulting leather will be hard and not supple.

10. If your end product is hard, just do it again with a second egg mix.

11. Once your hide is soft and supple, it is finished but not weatherproof. If you use a smoker and coat the fur with smoke it will shed water better. I skip this part as I am not going forth in the mountains with rabbit hide underwear as my childhood hero Sam Gribley did.

10

EQUIPMENT TO MAKE BUTCHERING EASIER

How to Make a Neck Wringer

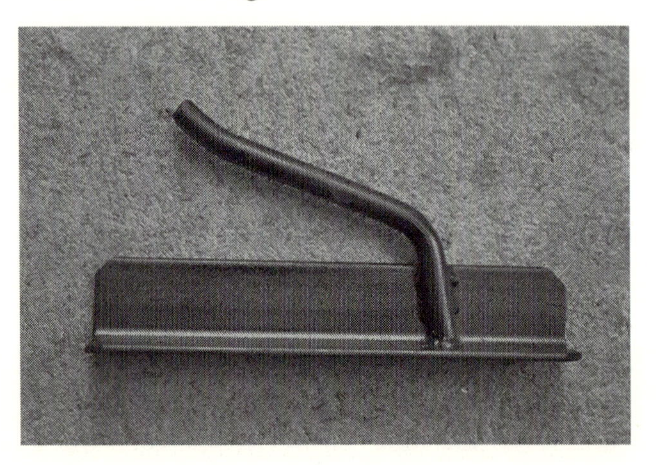

I use cervical dislocation as my preferred method of rabbit dispatching. When done properly it is fast and painless. It is actually considered by most veterinary organizations as one of the most ethical methods for terminating the life of small rodents such as rats, mice, squirrels. As a matter of fact there was a time that cervical dislocation was considered the most ethical method of terminating American convicts.

However, it takes quite a bit of strength and some measure of technique to perform cervical dislocation with your hands only. In the Wild West, they used stiff rope and a calculated drop to dislocate convicted murderers. That's not really appropriate with rabbits, but there are some devices on the market to help with rabbits.

There is a company that builds a device that helps with internal decapitation of rabbit. The build a stainless steel rack that the rabbits head fits into; this allows the farmer to use both hands to put sufficient pressure on the rabbit's legs instead on only one hand like manual methods use. I wanted to buy one, but it would not fit where I wanted it too because of its length, and I wanted a project I could do with my nephew while he was visiting (every 11 year old needs to spend a little time heating and banging on metal).

All I did was go to the building supply store and buy some metal rod and some mild steel strap.

I heated the rod with a torch and bent it into a "V" shape with a leather glove and a vice.

Next I welded one end of the "V" to a section of metal strap.

I drilled some mounting holes in the strap and nailed it to a fence post.

I did not make a plan, I just eyeballed everything. It works well and I have a video in the links section at the end of this document.

If you can't weld you can buy a commercially made rabbit wringer for around $75 dollars. I have no business with any company making these devices, but in the interest of supplying information for those that can't weld you can find such a device at rabbitwringer.com

To use a device like this you place the head into the “V” and pull back and up on the legs, dislocating the cervical spine. It is the quickest and least painful method of dispatching rabbits I have found. Once again, I have a video in the links section to walk you through the process.

How to Build a Small Game Gambrel

I have had trouble with butchering my rabbits as I did not have a proper Small Game Gambrel to spread the legs and keep the carcasses where they needed to be.

Luckily in discussing it with my friend, he showed me a picture of a commercial Small Game Gambrel, and between the two of us we were able to modify it to my use.

I have a video on the process in the links to videos section in rear of this document.

Building a small game gambrel was pretty easy, it just took some aircraft cable, some crimp ends, and a length of PVC pipe.

By threading the wire through the pipe the weight of the animal pulls the gambrel tightly on the legs so it is a very sturdy setup.

Once the animal is cleaned and ready to be removed, it is a simple matter of either cutting the feet off, or lifting it up to remove the weight from the hoops.

DIY Battery Powered Camp Sink

If you are going to butcher chickens, you need a sink. My wife doesn't like it when I do it her sink, so I 'made' my own. This is section is identical to the section found in my Chicken Basics book, so if you have that you can skip this part.

The Sink

Camp Sink

I saw this portable fish processing table and thought it would be a great start to make a battery powered camp sink. Now normally I would have bought a table and cut out a hole for a large Tupperware tote and rigged up a sink faucet, but I decided to try to make something nice this time.

This $65.99 table is designed to hook to a water hose, but I wanted to make it independent of a water hook up. You know just in case I wanted to butcher chickens in the wild outdoors.

When I can't find a solution on the internet my semi-modern brain starts to think that others tried my idea and found it unworkable. However, I am stubbornly optimistic and decided the lack of information online meant that I am innovative. After much scouring of the internet and all the Camping stores in Nashville, I decided on a small battery powered camp shower. I decided against a sump or boat bilge pump because of looks, but a handy person could definitely use something like that.

The Pump

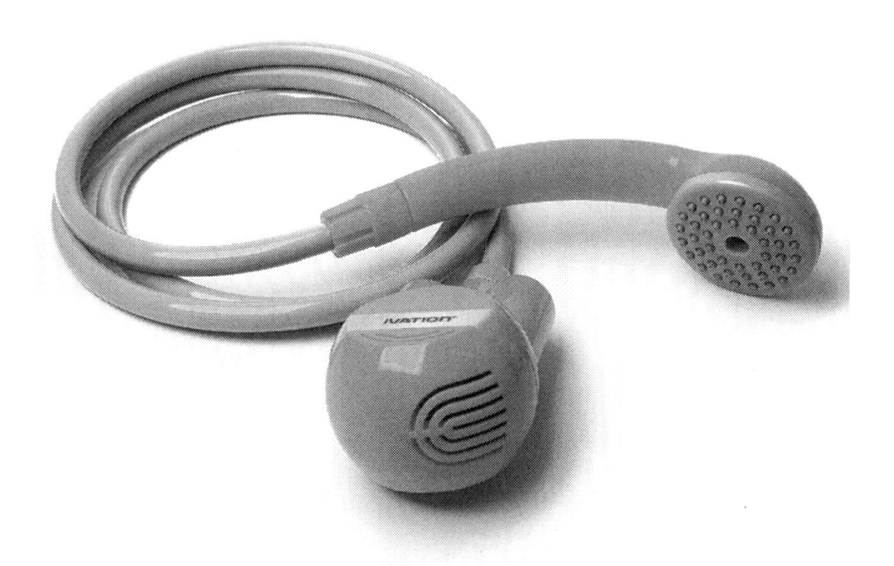

Battery Powered Shower

There are several battery powered showers online.

The shower head is 1/2 inch and is identical to the handheld shower heads you can buy at the store. Luckily the faucet on the processing table is also threaded the same way.

I went ahead and bought some brass fittings to go from 1/2 to 3/4 so I could plumb it to a garden hose because I did not open the box first.

You will also need two 5 gallon buckets. One for clean water, and one for dirty water.

Final Thoughts on the Camp Sink

For ease of use I tied a reused a mesh onion bag with a bar of soap in it to the side of the table, and bungee corded a roll of paper towels between the legs of the table.

Using this was pretty easy. All you need to do is press the on switch on the show pump and drop it in the clean water bucket. After a few seconds to prime you can turn on the faucet and have running water. Because this is battery powered, when not in use, you should remove the pump from the water. Additionally, the battery only lasts an hour or so on continuous running. For intermittent use, I got through all day Friday until early afternoon Saturday (with battery power to spare).

I like how this did not require any permanent modifications, so I can still use the shower as a shower. Which is important if I want my wife to go camping.

What I did not like about using the shower in my camp sink is that to turn on I had to press the button on the side of the pump. A newer version has the batteries and switch on a cord which I like much better. I will probably buy that later and attach the switch to the table itself.

Besides the inconvenience of having to turn it on and off, my only complaint is the depth of the sink. It really did not work well to wash pots. However, washing silverware and hands it was perfect.

If you were using this in your backyard to butcher chickens, hook it to a water hose and you will be much happier.

11

HOW TO BUTCHER RABBITS: A COMPLETE GUIDE

I like raising my rabbits. I enjoy watching them, feeding them, and knowing I am producing a good quality rabbit. That rabbit will become healthy meat for myself and my household. However, I don't enjoy the slaughtering portion of the meat process. If you are going to raise meat rabbits you need to know how to butcher rabbits.

I don't take joy from killing anything; however, if I am going to eat meat, an animal has to die. Personally, I feel that if I do it myself, I can avoid the waste and suffering that can come from an industrial process where efficiency is the most important aspect in slaughtering. I do my best to ensure the rabbit's life is the best it can be and its death is quick and painless as possible.

However, if you want to eat mean, somethings have to happen. This post shows how to butcher rabbits.

However, One thing is for sure, what my domestic rabbit suffers when I harvest it rabbit is nothing like the fate of a wild rabbits whose last moments is filled with running for its life, and then after being caught by its natural predators being ripped limb from limb while still alive.

Collect all your tools before you start.

Equipment:

• Very sharp skinning knife

• Heavy scissors

• 5-gallon bucket for water.

• A large square, possibly plastic, container

• Hose with nozzle

How to Kill the Rabbit

I spent a lot of time figuring out the best method for use to use to dispatch the rabbit. And while we will give you three other methods this is the method we use:

• Hold the rabbit by the hind feet using your non-dominant hand.

• With your dominant hand make a v with your thumb and index finger and wrap around the rabbits head so your palm is on the rabbits shoulder blades.

• Lift the rabbit's head up and backward while quickly pulling your hands apart. If you do it properly you will dislocate the head from the vertebral column, instantly killing the rabbit. The rabbit immediately stiffens into death-throes. You will only do it wrong once. If you are hesitant, and do not pull hard enough the rabbit's death cry will sound like a child. Believe me, that is not a sound you want to hear twice.

Here are Main Three Rabbit Dispatching Methods:

- Hold the rabbit around the shoulders and neck from behind with the left hand. Put the heel of your right hand under the rabbit's lower jaw (chin). Push sharply up and back with the right hand. This kills the rabbit by dislocating the neck, and does not damage the meat. This is called chinning

- Shoot the rabbit with a powerful pellet gun (at least 1000 fps) or a .22lr
- Broomstick the rabbit by sitting the rabbit on the ground, placing a broomstick or length of pipe across the neck and near the ears. Place your foot on one end. Have a partner stand on the other end. Grab the rabbit's legs and pull up and toward the pipe which will also instantly dislocate the neck.

Once Dead, Immediately Bleed the Carcass

Hang the rabbit by the hind legs. You could use hooks, or a small gambrel made out of welding rod. I use slipknots on a piece of string.

Immediately cut the head off. I use large scissors, but if you use a knife, cut along the area of dislocation. Your knife will slip easily through the dislocation in the neck.

Allow the rabbit to bleed out. Since the rabbit is dead the blood will drip and not spurt, but it's best to have a bucket to catch everything.

I don't whack them with sticks, as my aim is unsure and I want to make sure I kill them with the first attempt.

How to Skin the Rabbit

- Snip the skin around the foot of the rabbit. Take care to only cut skin and not tendon, muscle or bone. The fur around the rabbit's foot should look like a sock.
- Pull the skin on the leg down and work the skin down to the groin.
- Repeat on other leg.
- Insert your finger(s) between skin and belly at the groin, until your fingers can be seen at the second cut at the other side of the belly.

- Removing fingers, insert knife into the same space and cut up and outward, cutting the pelt free of the groin area.
- Now you can pull the pelt downward, loosening it from the carcass with your fingers, as needed. You'll be able to pull downward as far as the tail.
- With the skin pulled downward, insert your fingers under the skin below the tail, loosening the pelt from the back beside the tail.
- Remove your fingers and insert the knife into the same space, and cut the pelt loose from the tail, I pull the skin away from the groin and cut through the tail bone, sex organs, and anus all in one swoop.
- Pull the pelt downward, freeing it from the meat with your fingers as needed, pulling it down and over the neck like you would a T-shirt. The pelt will be turned itself inside out as you pull it off.
- Loosen and release the pelt from the neck area. Since the head is gone, soon the pelt will be attached only at the front paws.
- Cut the pelt free of the paws. Cut on the skin side only and avoid cutting any fur if you plan to use the pelt.

How to Clean the Rabbit

- Pinch up the abdominal lining and make a cut just below the groin. Be sure to only cut the skin and not any guts.
- Insert two fingers into the cut. Pulling outward a bit, insert knife, blade down, between your fingers. Cut downward, sliding your fingers downward with the knife, keeping the innards away from the edge of the blade. You only want to cut the abdominal wall. Cut down to the ribcage. The innards will fall outward a bit.
- Locate the pelvic joint. You're looking for the cartilage that joins both sides of the pelvis at the midline under the groin organs. It is less than an inch long. Press the knife inward

along the length of the joint. The cartilage should cut fairly easily. Ideally, the blade will separate the pelvic joint without cutting through to the rectum, which is directly under the pelvic joint.

- Grasp both thighs and bend them backward. This will spread the severed pelvic joint. You will be able to see if you need to carefully cut any other tissues alongside the intestines.
- Separate the innards from the liver. Everything should be falling outward except for the stomach, which is attached to the esophagus. Cut the stomach free of tissues and blood vessels. If you leave it attached to the esophagus and pull snugly, you may be able to pull the esophagus free of the neck, and the whole works will fall into the bucket below the carcass.
- Remove the kidneys along with the surrounding fat. There's a membrane around all that fat. With a little care, you can pull it all out without leaving globs of fat in the carcass.
- Gather up the liver carefully in your hand and cut it free.
- You'll easily locate the gallbladder, a small sac filled with a lot, or just a little, green gall. You'll also easily spot the gallbladder duct attaching the gallbladder to the liver. Pinch the duct (not the gallbladder) between a thumb and finger and pull it free from the liver. Drop it into the bucket.

Very important:

Once you pinch the duct and pull, you must not release your pressure on the duct until discarding it. The gall is extremely bitter, and should you lacerate the gallbladder or allow any of the gall to spill, it will ruin anything it touches.

- Cut through the rib cage close to one side of the breastbone. Spread the ribcage and cut the rabbit's diaphragm.
- Pull the lungs and the heart out of the chest cavity. Hopefully the trachea will come with them.

- Pull (or cut) the heart away from the lungs.
- Use a strong shears to cut off the front paws. Finish cutting any remaining tendons with the knife.
- Use the shears to cut the hind feet.

The whole butchering process should take between 10 – 15 minutes.

I save the liver, I would save the pelt, but at the young age of the rabbits, the skin is too thin to be used for most crafts.

The rest of the innards can be given to your dog or cat (or chickens).

How to Cut up the Carcass:

You'll get 8 rabbit pieces by following these directions: two front legs, two back legs, two rib sections, and two back sections.

- Separate the front limbs from the rib cage.
- Separate the hind limbs from the back
- Cut through the back-strap to separate the rib section from the back. With the meat cut, snap and break the back, dislocating it at the cut. Then it is easy to cut free.
- Bend the ribcage outward, and cut into its two sides.
- Cut the last strip, the back, down the middle. Cut the muscle and snap the back in two in order to cut through the joint.

Soak the carcasses in a sink full of cold salted water (2 tablespoons per gallon) for about a half-hour. This removes body temperature and helps dissipate the blood from the carcass. Rinse and either cook, or package for storage.

Now you see how to butcher rabbits, at least how I do it.

12

HOW TO CAN RABBIT MEAT

Canning rabbit is very similar to the way you would pressure can any meat.

Generally rabbit and chicken recipes are interchangeable. The main difference is that a rabbit has much less fat than a chicken. You need to keep that in mind for recipes when you are cooking a fresh rabbit because if you cook rabbit too fast the meat will end up tough and stringy.

So unless you are frying it, try and use the slower methods when cooking rabbit.

One great thing about pressure canning rabbit meat is you don't have to worry about tough meat.

That's because the meat was already pressured cooked and is very tender and moist. Canned rabbit is a tremendous time saver. As well as a good prep because you don't have to worry about powering your freezer.

I use canned rabbit in salads, casseroles, barbecue, in white gravies and sauces over biscuits and in any recipe that calls for cooked chicken.

While this post is about canning rabbit, and its part of a rabbit husbandry series, don't get too static in your thinking, you can interchange meats in canning recipes just as long as you remember to process the jars according to the ingredient that requires the longest processing time. And if you are canning chicken the times are the same.

I especially like white chili made with rabbit and is easily canned for me to take to work and keep in my desk. (Yes, my coworkers think I am crazy and wonder aloud at least once a week when I am going to eat the pint jar of turkey and rice I canned back in 2009 that I use as a discussion starter.)

Hot Pack or Raw Pack

Like the beef canning post you have a choice between the "hot pack" or "raw pack methods" as well as "bone in" or "bone out".

The quickest and (IMHO) best way in the long run and I think best way to can rabbit is with the hot pack, bone out method. This makes a canned product that is ready to go from the shelf – just like commercial canned meat.

A consideration for you to think about is that the flavor of the meat is stronger if you choose a bone in method, but its not off putting (to me) it's a subtle difference but the bones do add kind of a dark meat flavor.

The bone in method is simpler. Especially with small game animals. Squirrel and rabbit are hard to bone. (YES I said squirrel, it all eats the same)

Since there is no such thing as a free lunch. If you bone when you can, you can just heat and serve. If you don't bone when you can, you have to do it later.

Like the earlier post, I like using wide mouth jars; it's easier to pack, easier to get my meat out, and much easier to clean later.

As a general rule of thumb, allow 2 to 2 ½ pounds of boneless meat per quart. If your making bone-in canned rabbit you should allow for 2 ½ to 4 ½ pounds of meat per quart.

A Word About Giblets

If you are processing a large batch of rabbits and want to can the heart or livers (or if doing chickens the gizzards) set them aside to be canned in separate jars.

It's also a good idea to can the livers in their own jar because the liver taste will transfer to the other giblets.

If you have never canned meat before the video below will show you the process to can some rabbit, but if you have canned meat before just remember to be clean, follow your canner's instructions, wipe your jars, and be safe.

While I won't go into the step by step process since I have covered that before, I will tell you the processing times for small game and poultry:

Procedure:

Choose freshly killed and dressed, heathy animals. Large chickens are more flavorful than fryers. Dressed chicken should be chilled for 6 to 12 hours before canning. Dressed rabbits should be soaked 1 hour in water containing 1 tablespoon of salt per quart, and then rinsed. Remove excess fat. Cut the chicken or rabbit into suitable sizes for canning. Can with or without bones. The hot pack is preferred for best liquid cover and quality during storage. Natural poultry fat and juices are usually not enough to cover the meat in raw packs.

Hot pack

Boil, steam or bake meat until about two-thirds done. Add 1 teaspoon salt per quart to the jar, if desired. Fill jars with pieces and hot broth, leaving 1-1/4 inch headspace.

Raw pack

Add 1 teaspoon salt per quart, if desired. Fill jars loosely with raw meat pieces, leaving 1-1/4 inch headspace. Do not add liquid.

Adjust lids and process following the recommendations in the chart below

		Canner Pressure (PSI) at Altitude			
Jar Size	Process Time	0-2000 ft	2001 to 4000 feet	4001 to 6000 feet	6001 to 8000 feet
	Without Bones				
Pint	75 Minutes	11 lb.	12 lb.	13 lb.	14 lb.
Quart	90 Minutes	11 lb.	12 lb.	13 lb.	14 lb.
	With Bones				
Pint	65 Minutes	11 lb.	12 lb.	13 lb.	14 lb.
Quart	75 Minutes	11 lb.	12 lb.	13 lb.	14 lb.

PLEASE REVIEW

Please visit my Amazon Author Page at:

https://amazon.com/author/davidnash

If you like my work, you can really help me by publishing a review on Amazon.

The link to review this work at Amazon is:

https://www.amazon.com/review/create-review?asin=B07BPHHL8T

LINKS TO VIDEOS

The Basics of Raising Backyard Rabbits: Playlist

http://yt.vu/p/PLZH3jGjLQ0rCKG3oY_DgC96PcdnDU_vjS

Rabbit Husbandry

https://youtu.be/GRAEjnzkUsA

How to Tattoo Rabbits: Why You Should

https://youtu.be/R_lh6j-ANJ0

How to Bottle Feed Baby Rabbits

https://youtu.be/e2uhtRhQ4mw

How to Butcher Rabbits: A Complete Guide

https://youtu.be/QHTjr4syyXA

How to Build a Small Game Gambrel

https://youtu.be/ildd-FQSZbs

How to Make Rabbit Treat Sticks from Leftover Juicer Pulp

https://youtu.be/2lybCYUEeEs

How to Make a Homemade Rabbit Neck Wringer

https://youtu.be/HePwulgKSGM

How to Use a Rabbit Neck Wringer

https://youtu.be/d2MwgnYIIn8

How to Pressure Can Rabbit Meat

https://youtu.be/1wnmEL7MVts

Tarp Cover for Animal Cages

https://youtu.be/kG2Nk96fTJg

BONUS: RABBITRY RECORDS

Notebook containing records and logs essential to running a profitable rabbitry

Have you ever wanted to raise rabbits?

Do you raise rabbits but have a tough time keeping up with breeding dates?

Rabbitry Records is a compact notebook filled with the documents you need to take the stress out of raising rabbits.

It holds all the forms you need to run a stress free and profitable business.

Besides plenty of ruled note pages it contains:

- Buck Record Sheets
- Doe Record Sheets
- Rabbit Tattoo Logs
- Litter Growth Records
- Rabbits Purchased Records
- Rabbit Sales Records
- Feed and Expenses Sheets
- Misc. Income Sheets
- Income/Expense Summaries
- Notes

There is plenty of help just inside this book

Over the next few pages see some of the forms listed in this book, in the print notebook the forms fill the entire page.

Rabbitry Records for:

____

From: ____________________

To: ______________________

Includes:

- Buck Record Sheets (3)
- Doe Record Sheets (15)
- Rabbit Tattoo Log (2)
- Litter Growth Record (16)
- Rabbits Purchased Record (2)
- Rabbit Sales Record (16)
- Feed and Expenses Sheet (10)
- Misc. Income Sheet (3)
- Income/Expense Summary (2)
- Notes (10)

____

Rabbit Breeding Record (Doe)

Name:________________________ DOB:____/____/_____ Weight:__________ Breed:____________________

Ear #:_______ Sire: ___________________Dam: ____________________ Fertile Date: _____________________

Breed Date	Buck	Nest Box Date (28)	Due Date (31)	Kindled Date	# Born Alive	# Born Dead	X Kits Weaned	8 Week Weight	Comments

Litter Growth Record

Litter ID #	Sire	Dam	4 Week	6 Week	8 Week	10 Week	12 Week	14 Week	16 Week

Rabbit Sales Record

Date	Tattoo #	Description	Ped ?	Price	Buyer Name	Phone	Email	City State

Feed & Misc. Expenses

Date	Description	Quantity	Total Cost

_______ Income/Expense Summary
Year

Income	
Rabbit Sales	
Miscellaneous Income	
Total Income	
Expenses	
Rabbits Purchases	
Feed Expenses	
Miscellaneous Expenses	
Total Expenses	
Financial Summary (Total Income – Total Expenses	
Circle One	**Profit Loss**

Notes:

____

____

____

____

____

____

____

____

____

BONUS: EXCERPT FROM THE BASICS OF BACKYARD BEEKEEPING

Why Keep Bees

Personally, I had wanted to get bees for a long time, but I was afraid it was too hard, or I would get stung too much.

After a little research and seeing how important bees are to agriculture and how valuable they are I decided to try my hand at beekeeping. For me it has been a very fruitful journey and I haven't been stung so much that the honey and wax have not made it worthwhile

In addition to using honey on my peanut butter sandwiches, Honey is also a great replacement for sugar, useful for wound management, seasonal allergy prevention, as well as using the pollination of the bees to increase my garden yield.

It stores indefinitely, and I found out, you can use it to make alcoholic beverages.

Besides, bees are indispensable to modern agriculture. Without bees to pollinate crops, we starve to death.

Plusses everywhere, beekeeping sounded to me like a win-win situation. The only think keeping me back from keeping bees was my wallet, and my relationship with my new neighbors.

When I started beekeeping all I knew was that there are two types of hives, the traditional langstroth hive everyone is familiar with (white boxy thing on a concrete block) and a top bar hive. (cheaper and easier but produces less honey and more wax as well as taking more knowledge).

I also knew that in my state, Tennessee, you are required to register and get a bee keeping license. The license was free and I applied, even though I cannot get my new bees until the spring. I will spent that winter loading up on all the toys, building my hives, reading, and most importantly, making sure my wonderful wife is on board with another project.

The following chapters cover all I learned in that year and the following 7 years as I became a better beekeeper.

If you like this Introduction to The Basics of Backyard Beekeeping, you can find it on Amazon.

ALSO BY DAVID NASH

Homestead Basics

The Basics of Raising Backyard Chickens

The Basics of Raising Backyard Rabbits

The Basics of Beginning Beekeeping

The Basics of Homemade Cheesemaking

The Basics of Making Homemade Wine and Vinegar

The Basics of Homemade Cleaning Supplies

The Basics of Baking

The Basics of Food Preservation

The Basics of Food Storage

The Basics of Cooking Meat

The Basics of Make Ahead Mixes

The Basics of Beginning Leatherwork

Note and Record Books

Rabbitry Records

Correction Officer's Notebook

Get Healthy Notebook

Non Fiction

21 Days to Basic Preparedness

52 Prepper Projects

52 Prepper Projects for Parents and Kids

52 Unique Techniques for Stocking Food for Preppers

Basic Survival: A Beginner's Guide

Building a Get Home Bag

Handguns for Self Defense

How I Built a Ferrocement "Boulder Bunker"

New Instructor Survival Guide

The Prepper's Guide to Foraging

The Prepper's Guide to Foraging: Revised 2nd Edition

The Ultimate Guide to Pepper Spray

Understanding the Use of Handguns for Self Defense

Fiction

The Deserter: Legion Chronicles Book 1

The Revolution: Legion Chronicles Book 2

Collections and Box Sets

Preparedness Collection

Translations

La Guía Definitiva Para El Spray De Pimienta

Multimedia

Alternative Energy

Firearm Manuals

Military Manuals 2 Disk Set

ABOUT THE AUTHOR

David Nash is a suburban homesteader with chickens, bees, rabbits, and a couple of goats in his suburban yard. For a while he even had an extensive aquaponics setup in his basement, until his long-suffering wife made him eat all the fish.

He knows how to raise animals humanely, simply, and without angering the neighbors. Dave runs a popular YouTube channel on DIY homesteading as well as being the author of several books on DIY preparedness and urban homesteading topics.

In fact, the tips shown in this book contributed to him receiving the third highest preparedness score on the TV show Doomsday Preppers

He is a father and a husband. He enjoys time with his young son William Tell and his school teacher wife Genny. When not working, writing, creating content for YouTube, playing on his self-reliance blog, or smoking award-winning BBQ he is asleep.

amazon.com/author/davidnash
facebook.com/booksbynash
youtube.com//tngun
goodreads.com/david_allen_nash
twitter.com/dnash1974
instagram.com/shepherdschool
pinterest.com/tngun

Made in the USA
Middletown, DE
31 March 2021

36593946R00054